A Picture Book of

DESERT ANIMALS

Written by Joanne Gise
Illustrated by Roseanna Pistolesi

W9-CUH-610

Troll Associates

CACTUS WREN

The prickly cactus plant may not seem like a friendly place to live, but the cactus wren thinks it's great! These wrens like to eat the cactus' fruit and seeds. They also like to eat insects and lizards that run along the ground under the cactus.

The cactus protects the wren's nest, too. The bird builds her nest between the cactus' thorny spines. Any animal that might want to eat the baby birds will think twice before trying to get past those sharp spikes!

After the cactus wren lines the nest with feathers, she lays 3-7 eggs in it. These eggs hatch in about 16 days. The wren may lay 2 or 3 *broods*, or groups of eggs, each spring.

Library of Congress Cataloging-in-Publication Data

Gise, Joanne.
    A picture book of desert animals / by Joanne Gise; illustrated by Roscanna Pistolesi.
        p.   cm.
    Summary: Describes a variety of desert animals, including the cactus wren, kangaroo rat, and scorpion.
    ISBN 0-8167-2148-3 (lib. bdg.)   ISBN 0-8167-2149-1 (pbk.)
    1. Desert fauna—Juvenile literature.   2. Desert fauna—Pictorial works—Juvenile literature.   [1. Desert animals.]   I. Pistolesi, Roscanna, ill.   II. Title.
QL116.G57   1991
591.909'54—dc20   90-40436

ANTELOPE JACK RABBIT

A jack rabbit isn't a rabbit at all. It's a *hare*. Hares and rabbits are related, but they are different from each other in several important ways. A hare's young are born with fur and with their eyes open. Baby rabbits are blind and hairless when they are born. Also, if a rabbit senses danger, it usually sits very still and tries to hide. A hare will use its long, strong back legs to leap away.

The antelope jack rabbit is the largest kind of jack rabbit. It weighs about 8 pounds and is over 2 feet long. It likes to eat the leaves and stems of juicy plants. The water in these plants helps the jack rabbit survive in its dry home.

KANGAROO RAT

This cute little animal got its name because it can jump long distances, just like a kangaroo does. A kangaroo rat's tail is about as long as the rest of its body. This helps the rat keep its balance when it jumps.

Kangaroo rats are *nocturnal*. This means they are active during the night. During the day, it is very hot in the desert. The kangaroo rat stays out of the heat by sleeping in its underground burrow. Here it is much cooler.

Kangaroo rats do not need to drink any water! They get all the water they need from the plants and seeds they eat.

COYOTE

The quiet of the desert night is suddenly broken by a spooky howl. What makes such an eerie sound? It is the coyote (ky-OH-tee or KY-oht). These animals might be talking to one another, or just ''singing'' for the fun of it.

Coyotes are related to wolves, foxes, and dogs. They like to eat hares, mice, squirrels, insects, fruit, and even big animals, such as sheep or antelope.

Coyotes usually live alone or in pairs. In the spring, the female gives birth to 5 or 6 *pups*. Baby coyotes are very tiny when they are born, and they are blind for the first 2 weeks of their lives. But by late summer, they can take care of themselves. Then it's time to leave their parents and go out on their own.

ROADRUNNER

The roadrunner is very well named.
Although it can fly, this bird loves to run.
It can move as fast as 15 miles an hour on its long, thin legs.
This speedy way to travel helps the roadrunner catch the
insects, lizards, mice, snakes, and small birds it likes to eat.

Male and female roadrunners live together all year long. In
the spring, they build a cup-shaped nest out of twigs and line it
with soft leaves and grass. The female lays her eggs here. After
2 or 3 weeks, 2-6 very hungry little chicks hatch. Their mother
and father take turns feeding and caring for them until the
chicks are about 2 weeks old. Then the young birds can look
after themselves.

HORNED LIZARD

You may have heard someone call this odd-looking creature a *horned toad*. It got that name because its body is shaped a little like a toad's. But this animal is really part of the lizard family.

A horned lizard's body is covered with sharp spikes. It even has spikes on the back of its head! These protect it from attackers. A horned lizard can also squirt blood from its eyes to scare away other animals.

KIT FOX

When you look at a kit fox, the first thing you see is ears. This little animal's large ears almost seem too big for its head. But there is an important reason for those big ears. They are full of blood vessels. If the fox gets too hot, its body heat can escape through these blood vessels into the air.

Kit foxes live in underground tunnels called *burrows*. They usually stay in these tunnels all day. At night, when it is cooler, they come out to hunt insects, mice, and other small animals.

GILA MONSTER

The Gila (HEE-luh) monster isn't really a monster. It is an animal called a *lizard*. Lizards are *cold-blooded*. This means their body temperature is the same as the air around them. If a lizard stays out in the desert sun for too long, its body can become so hot that it will die. So lizards like to spend part of their time in the shade of a rock or a bush.

A Gila monster's stubby tail is full of fat. The lizard can go for months without eating. It lives off the fat in its tail. When it does eat, a Gila monster thinks bird and snake eggs and small animals are tasty.

A Gila monster is shy and likes to be left alone. But if something bothers it—watch out! Gila monsters are poisonous. Their bite is very painful, and it can kill some animals.

SCORPION

The scorpion may be small—it is no more than 8 inches long—but it can be very dangerous. At the end of its tail is a curved stinger. Two glands at the bottom of the stinger are full of poison. A scorpion's sting can make a person sick, and it can kill many animals.

Scorpions usually come out at night to eat insects and spiders. A scorpion's coloring helps it hide against the sand and rocks of its home, so its prey cannot see it coming.

ELF OWL

A cactus makes a perfect home for the tiny elf owl, which is only about 5 inches long. The owl waits for a bird called the desert wood-pecker to move out of the hole it has dug inside a cactus. Then the elf owl moves in.

During the day, the owl sleeps in its nest. It flies out into the night to hunt insects, scorpions, and small snakes and lizards.

ANTELOPE GROUND SQUIRREL

Most squirrels live in trees. But the ground squirrel lives— you guessed it!—under the ground. These animals build their nests in underground burrows. This protects them from the hot desert sun. It also helps them hide from animals like coyotes, foxes, and birds. They think a squirrel makes a good snack!

TURKEY VULTURE

These birds perform a very important job. They are the garbage collectors of the desert. They fly high in the sky, searching for dead animals to eat. A turkey vulture's very good sense of smell helps it find its food.

Turkey vultures are quite big. They are about 2½ feet long and their wingspan can be up to 6 feet wide. They live in many places, from the tip of South America all the way up to southern California.

23

SIDEWINDER

A sidewinder leaves very interesting tracks as it moves along. Keeping its head and tail on the ground, this snake lifts its body sideways. When the body touches the ground, the head and tail follow it. This strange movement helps the snake in many ways. It lets very little of the snake's body touch the hot ground at one time, and it keeps the snake from sinking into the soft, loose sand. It also helps the snake sneak up on mice and other animals it wants to gulp down for dinner.